Atom to steel

Interview chain questions hack

———

Bineeth Benny

Atom to steel

Copyright © by Bineeth Benny

All rights reserved. No part of this book may be reproduced or transmitted in any form or by any means without written permission from the author.

Dedication

I dedicate this work to mechanical engineers attending interviews in material science field.

Email: expertvineeth@outlook.com

Chapter one Basics..06
Chapter two cells... 10
Chapter three crystal structure....................... 13
Chapter four phases..15
Chapter five Iron carbon diagram ....….....17
Chapter six Heat treatments 26
Chapter seven Misallaneous ....…............ 29

Preface

Mechanical engineering basics related to material science was searched online for each answer I found a chain question which may be asked by the interviewer. I have taken keen interest in finding the best and simple answer from the internet sources to make this book.

Chapter One Basics

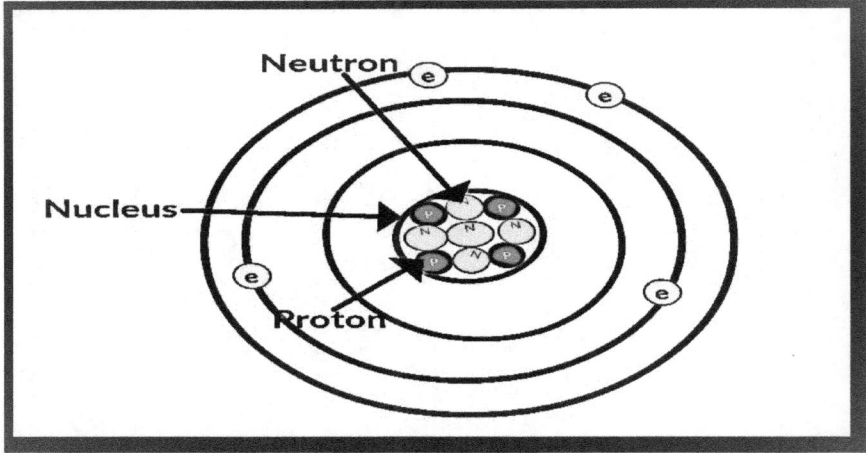

Definitions

1) Atoms are the basic building blocks of ordinary matter. Atoms can join together to form molecules, which in turn form most of the objects around you. Atoms are composed of particles called protons, electrons and neutrons.

2) Proton: is a subatomic particle found in the nucleus of every atom. The particle has a positive electrical charge, equal and opposite to that of the electron. If isolated, a single proton would have a mass of only 1.673 10^{-27} kilogram, just slightly less than the mass of a neutron.

3) Electron: is a negatively charged subatomic particle. It can be either free (not attached to any atom), or bound to the nucleus of an atom. Electrons in atoms exist in spherical shells of various radii, representing energy levels. The larger the spherical shell, the higher the energy contained in the electron.

3.1) Valence Electron: The electrons in the outermost shell are the valence electrons--the electrons on an atom that can be gained or lost in a chemical reaction. Since filled d or f subshells are seldom disturbed in a chemical reaction, we can define valence electrons as follows: The electrons on an atom that are not present in the previous rare gas, ignoring filled d or f subshells.

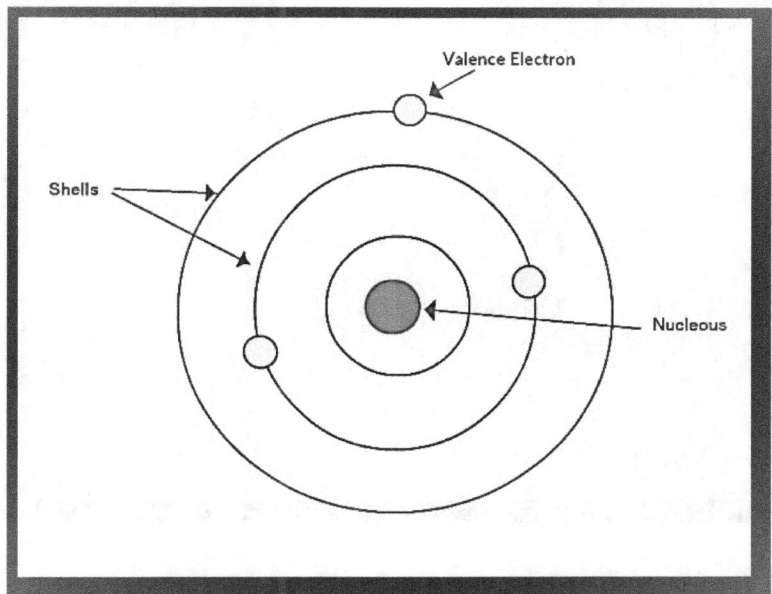

3.2) Shells: here the electrons generally stay. There are 4 types of electron shells: s, p, d and f shells.

4) Neutron: A neutron is one of the subatomic particles that make up matter. In the universe, neutrons are abundant, making up more than half of all visible matter. It has no electric charge and a rest mass equal to 1.67493×10^{-27} kg—marginally greater than that of the proton but nearly 1839 times greater than that of the electron.

5) Nucleus: is the center of an atom. It is made up of nucleons (protons and neutrons) and is surrounded by the electron cloud. The size (diameter) of the nucleus is between 1.6 fm (10^{-15} m) (for a proton in light hydrogen) to about 15 fm (for the heaviest atoms, such as uranium).

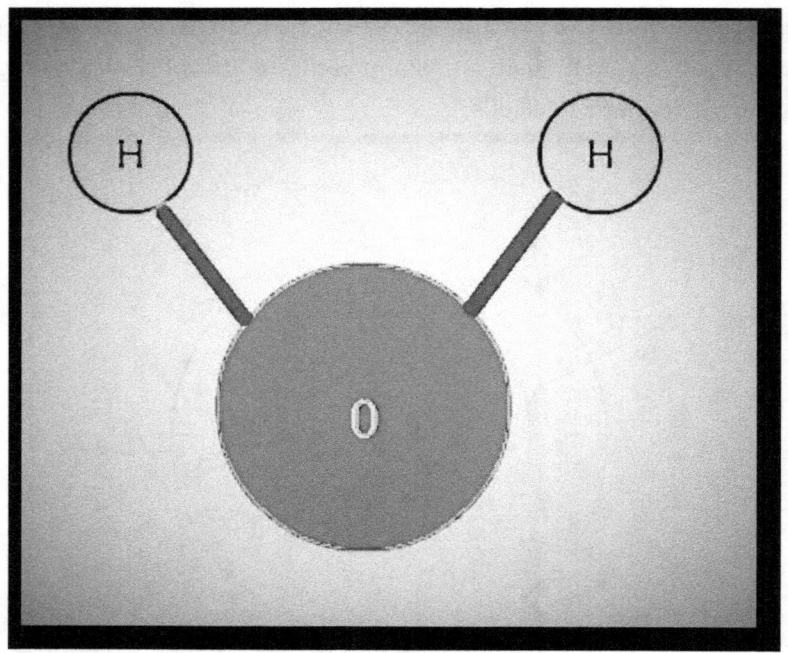

6) Molecule: A molecule is formed when two or more atoms join together chemically.

7) Compound: A Compound is a molecule that contains at least two different elements. All compounds are molecules but not all molecules are compounds.

8) An ion: Atom or molecule having a positive charge (is an 'anion') or a negative charge (is a 'cation') due to loss or gain of electrons during a chemical action or exposure to some types of radiation. Ions behave very differently from the atoms or molecules they are formed from.

9) Types of Bond: 9.1) Ionic Bond, 9.2) Covalent bond, 9.3) Polar bond and 9.4) Hydrogen bond.

9.1) Ionic Bond: Ionic bonding is the complete transfer of valence electron(s) between atoms. It is a type of chemical bond that generates two oppositely charged ions. In ionic bonds, the metal loses electrons to become a positively charged cation, whereas the nonmetal accepts those electrons to become a negatively charged anion. Ionic bonds require an electron donor, often a metal, and an electron acceptor, a nonmetal.

Metals loose or gain electrons, to achieve noble gas configuration and satisfy the octet rule. Similarly, nonmetals that have close to 8 electrons in their valence shells tend to readily accept electrons to achieve noble gas configuration.

(Greed makes you Negative # Hence loose electron and Become positive)

9.2) Covalent bond: is the interatomic linkage that results from the sharing pair of electrons between two atoms. The binding arises from the electrostatic attraction of their respective nuclei (Nucleus) for the same electrons. A covalent bond forms when the bonded (Unity) atoms have a lower total energy than that of widely separated atoms (individual).

(Sharing of common things # ☐ Loss of Strength in unity # Covalent)

9.3) Polar bond: A water molecule, H2O, is an example of a polar covalent bond. The electrons are unequally shared, with the oxygen atom spending more time with electrons than the hydrogen atoms. Since electrons spend more time with the oxygen atom, it carries a partial negative charge.

9.4) Hydrogen bond: Hydrogen bonding is a force of attraction between slightly positive hydrogen of one molecule to a slightly negative region of another molecule. An individual hydrogen bond is not very strong.

Chapter Two CELLS

10.1) What is a unit cell?
Ans: The smallest part of a crystal is called as unit cell. It is formed by combination of atoms1 and molecules6.

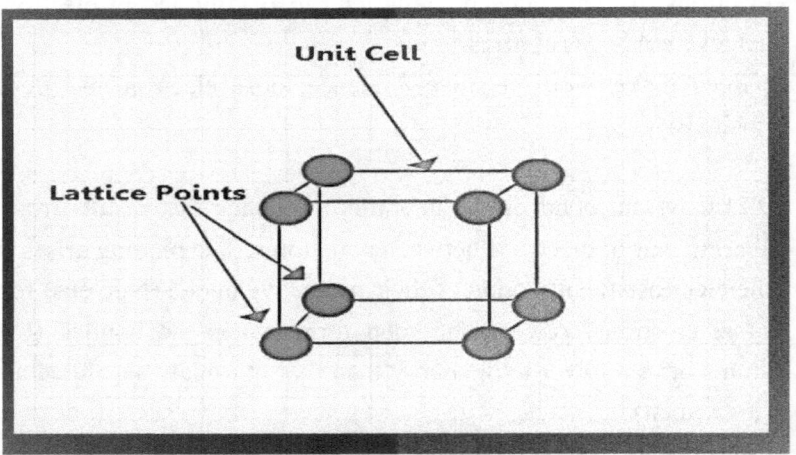

10.2) what is a crystal structure?
Ans: The crystal structure can be formed by the repetition of unit cells10.1.

10.3) what is grain, in a metal?
Ans: Metals10.4 have a crystalline structure 10.2 .When a metal solidifies from the molten state, millions of tiny crystals start to grow. The longer the metal takes to cool the larger the crystals grow. These crystals form the grains in the solid metal. Each grain is a distinct crystal with its own orientation.

10.3) what are grain boundaries?
Ans: The areas between the grains are known as grain boundaries.

10.4) what is metal? Define: Metals are opaque10.5, lustrous10.6 elements that are good conductors of heat and electricity. Most metals are malleable10.7 and ductile10.8 and are, in general, denser than the other elemental substances.

10.5) Opaque: not transparent.

10.6) Lustrous: shining.

10.7) Malleable (Malleability): able to be hammered or pressed into shape without breaking or cracking.

10.8) Ductile (Ductility): by virtue of this property, metal able to be drawn out into a thin wire.

10.9) Conductors (Conductivity): 10.9.1) Electrical conductivity in metals is a result of the movement of electrically charged particles.
The atoms of metal elements are characterized by the presence of valence electrons 3.1 - electrons in the outer shell of an atom that are free to move about. It is these 'free electrons' that allow metals to conduct an electric current. Because valence electrons are free to move they can travel through the lattice that forms the physical structure of a metal. The transfer of energy is strongest when there is little resistance.

10.9.2) Thermal Conductivity: Thermal conductivity, describes the transport of energy – in the form of heat – through a body of mass as the result of a temperature gradient. According to the second law of thermodynamics, heat always flows in the direction of the lower temperature.

10.4) What is electrical resistance of a metal? :Resistance is the opposition that a substance offers to the flow of electric current. It is represented by the letter R. The standard unit of resistance is the ohm, and symbolized by the Greek letter omega.

When an electric current of one ampere passes through a component across which a potential difference (voltage) of one volt exists, then the resistance of that component is one ohm.

Chapter Three CRYSTAL STRUCTURE

11.1) what are the basic cubic crystal structures?: Simple Cubic 11.2, Body-Centred Cubic (BCC)11.4 & Face-Centred Cubic (FCC)11.5.

11.2) what is simple cubic structure? Ans: simple cubic structure: a crystalline structure with a cubic unit cell with lattice points11.3 only on the corners.

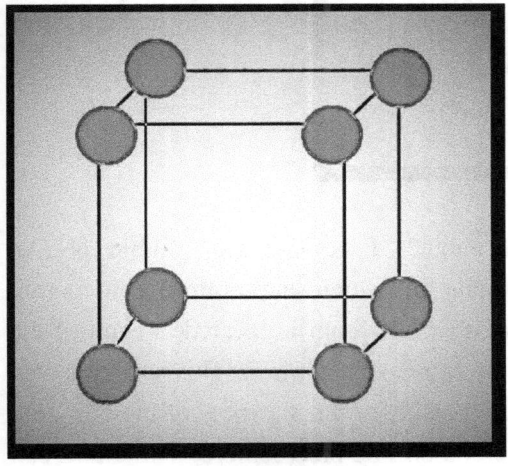

11.3) What are lattice points?: Lattice points are the positions (or coordinates if you wish) where you can place an atom. Lattice points can be vacant.

11.4) what is body centered cubic structure ?: The body-centered cubic unit cell has atoms at each of the eight corners of a cube (like the cubic unit cell) plus one atom in the center of the cube Each of the corner atoms is the corner of another cube so the corner atoms are shared among eight unit cells.

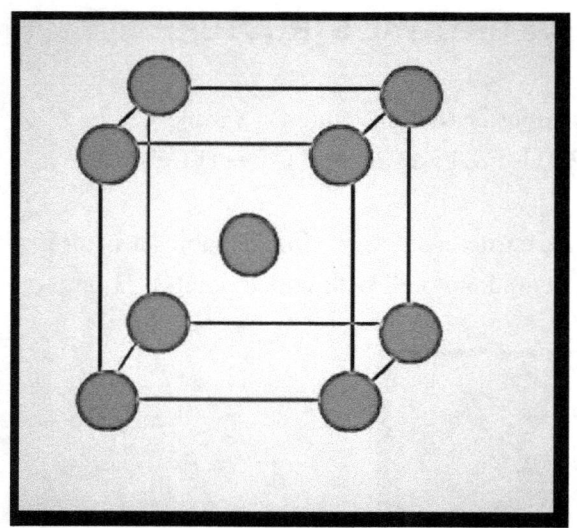

11.5) what is Face centered cubic?: Face Centered Cubic is ,An arrangement of atoms in crystals in which the atomic centers are disposed in space in such a way that one atom is located at each of the corners of the cube and one at the center of each face. This structure also contains the same particles in the centers of the six faces of the unit cell, for a total of 14 identical lattice points. The face-centered cubic unit cell is the simplest repeating unit in a cubic closest-packed structure.

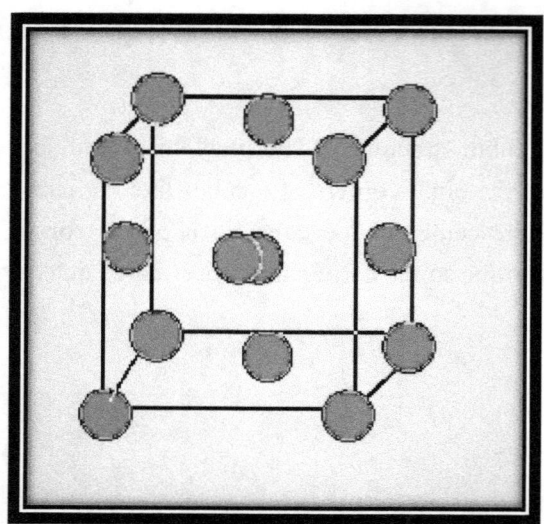

Chapter Four PHASES

12.1) what are phases? In metallurgy: Phase is used to refer to a physically homogeneous state of matter, where the phase has a certain chemical composition, and a distinct type of atomic bonding and arrangement of elements. Within an alloy, two or more different phases can be present at the same time.

12.2) what are the types of phases known? (Iron-carbon diagram): Ferrite12.3, Cementite12.4, Austenite12.5, Pearlite12.7 Bainite12.8 and Martensite12.9.

12.3) what is Ferrite?: Ferrite is a ceramic-like material with magnetic properties, which is used in many types of electronic devices. Ferrite is used in, Permanent magnets, Ferrite cores for transformers and toroidal inductors, Computer memory elements etc.

12.4) what is cementite?: The hard brittle compound of iron and carbon that forms in carbon steels and some cast irons. Formula: Fe_3C (Cited 1 A).

12.5) what is Austenite?: Austenite was originally used to describe an iron-carbon alloy, in which the iron was in the face-centred-cubic (gamma-iron) form. It is now a term used for all iron alloys with a basis of gamma-iron. Austenite in iron-carbon alloys is generally only evident above 723°C, and below 1500°C, depending on carbon content.

12.6) Can austenite be retained at room temperature?: yes, it can be retained to room temperature by alloy additions such as nickel or manganese.

12.6) Can austenite be retained at room temperature?: yes, it can be retained to room temperature by alloy additions such as nickel or manganese.

12.7) what is Pearlite?: Pearlite is frequently said to be two-phased, lamellar (layered or plate-like) structure composed of alternating layers of alpha-ferrite (88%) and Cementite (12%) that occurs in some steels and-cast-irons.

12.8) what is Bainite ?: Bainite is a microstructural crystalline pattern that forms in steel during heating , Austenite must be cooled rapidly enough so that pearlite does not form, Cooling in the austenite must also be delayed long enough to prevent martensite from forming.It possesses some of the extreme hardness of martensite, as well as the tough structure-of-pearlite.

12.9) what is Martensite : Martensite, named after the German metallurgist Adolf Martens (1850–1914), is any crystal structure that is formed by displacive transformation, as opposed to much slower diffusive transformations.Martensite is the kinetic product of rapid cooling of steel containing sufficient carbon.it is not shown in the equilibrium phase diagram of the iron-carbon system because it is a metastable phase.

Chapter Five Iron Carbon Diagram. (Brief Explanation)

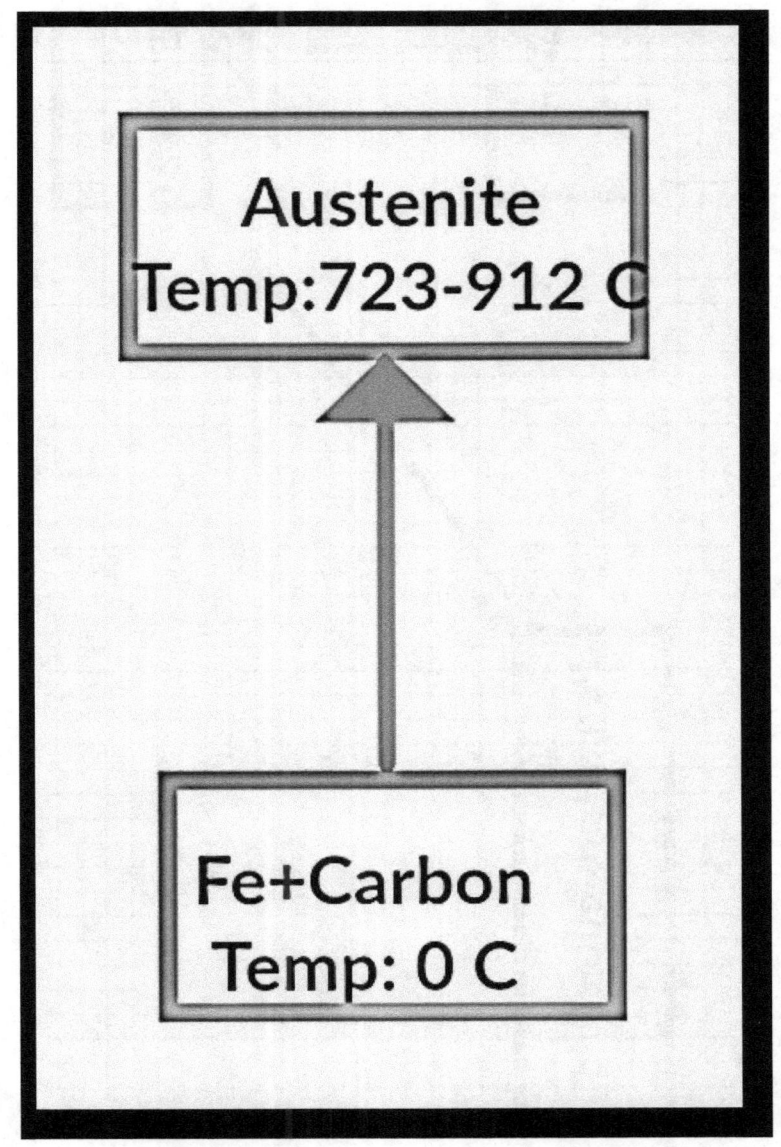

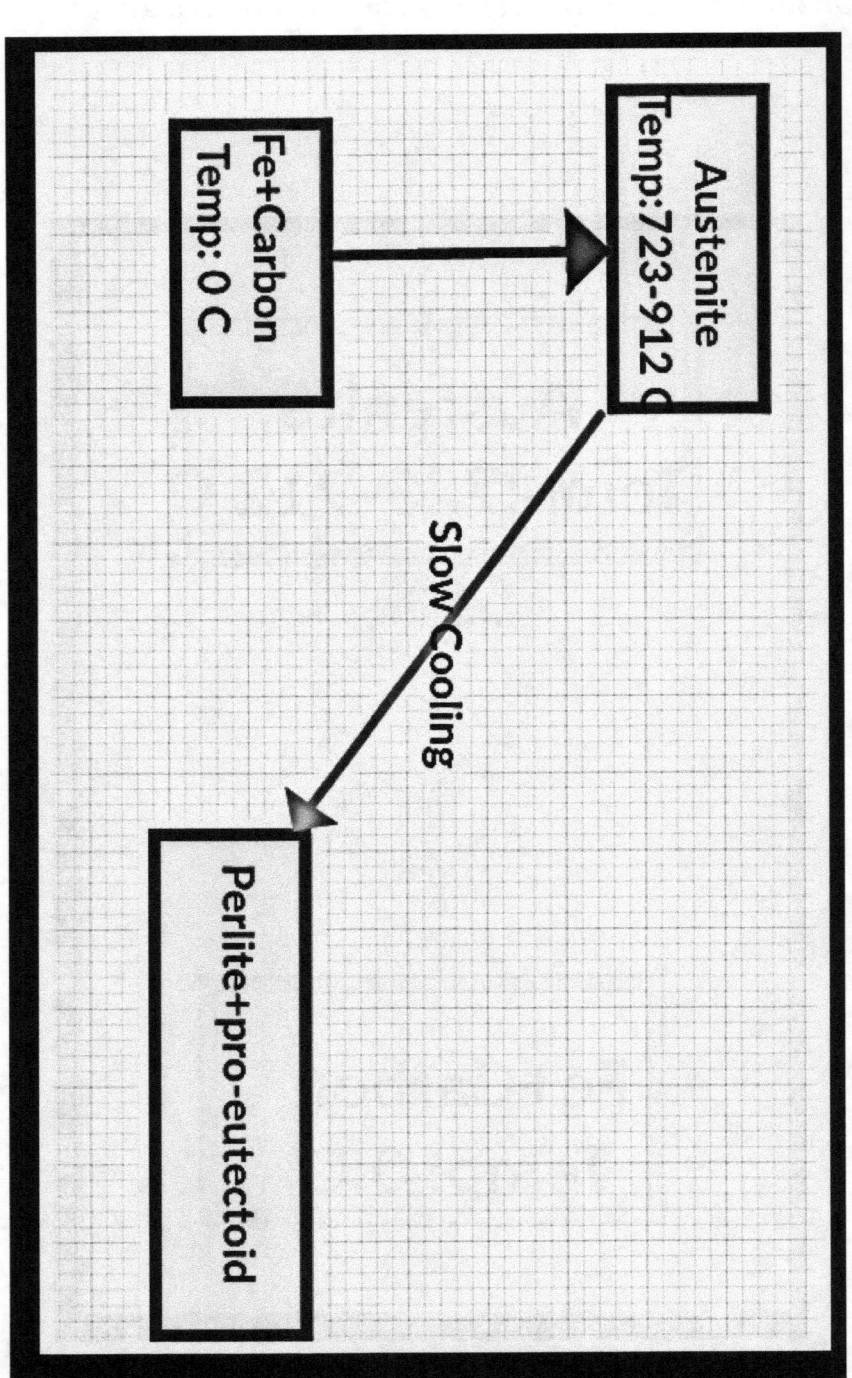

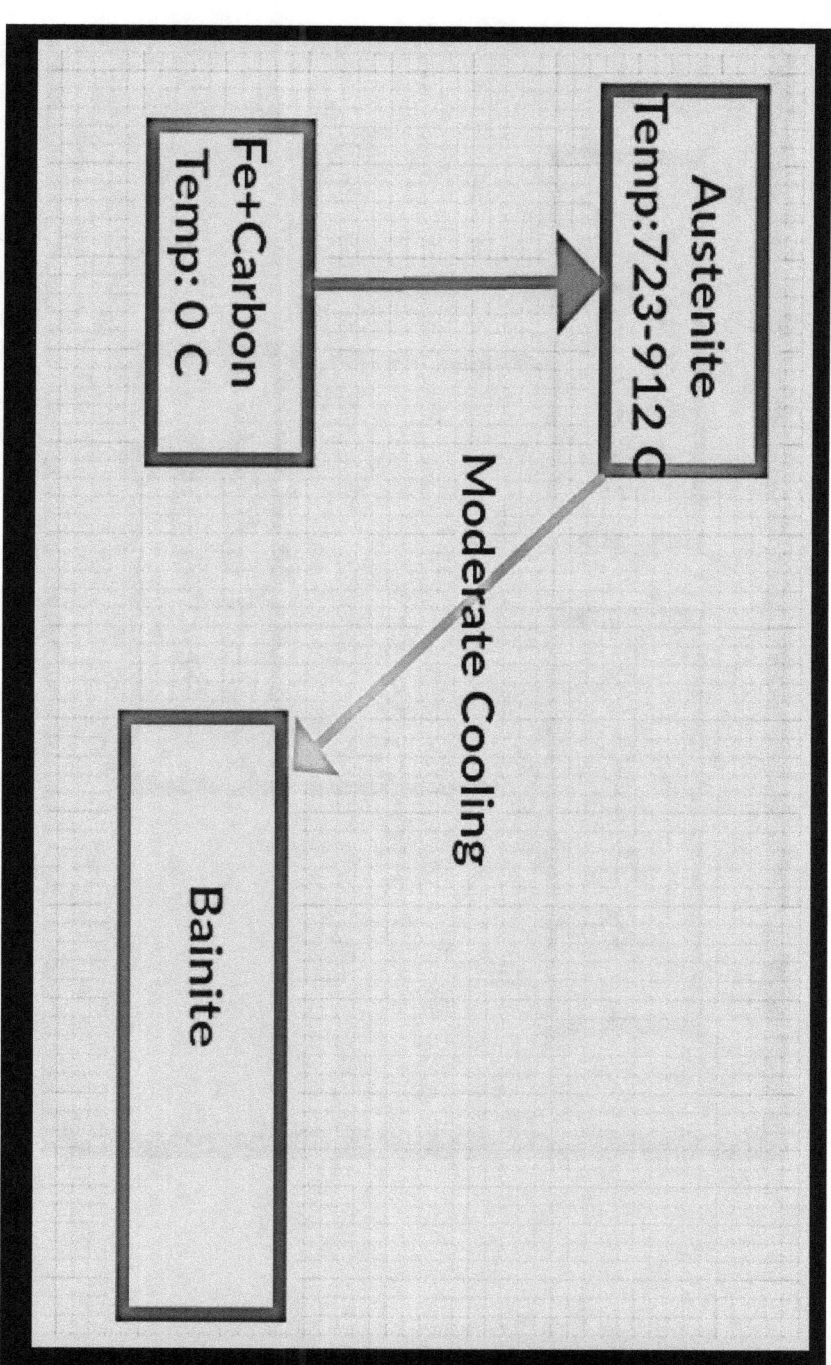

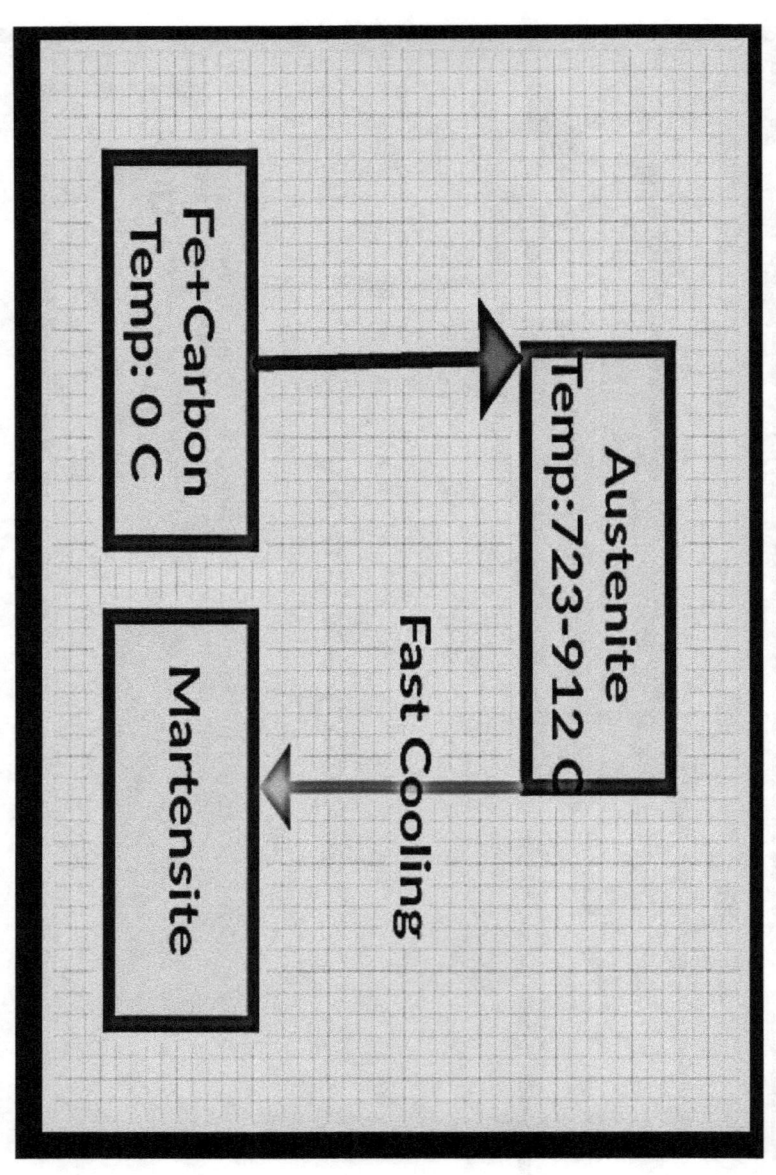

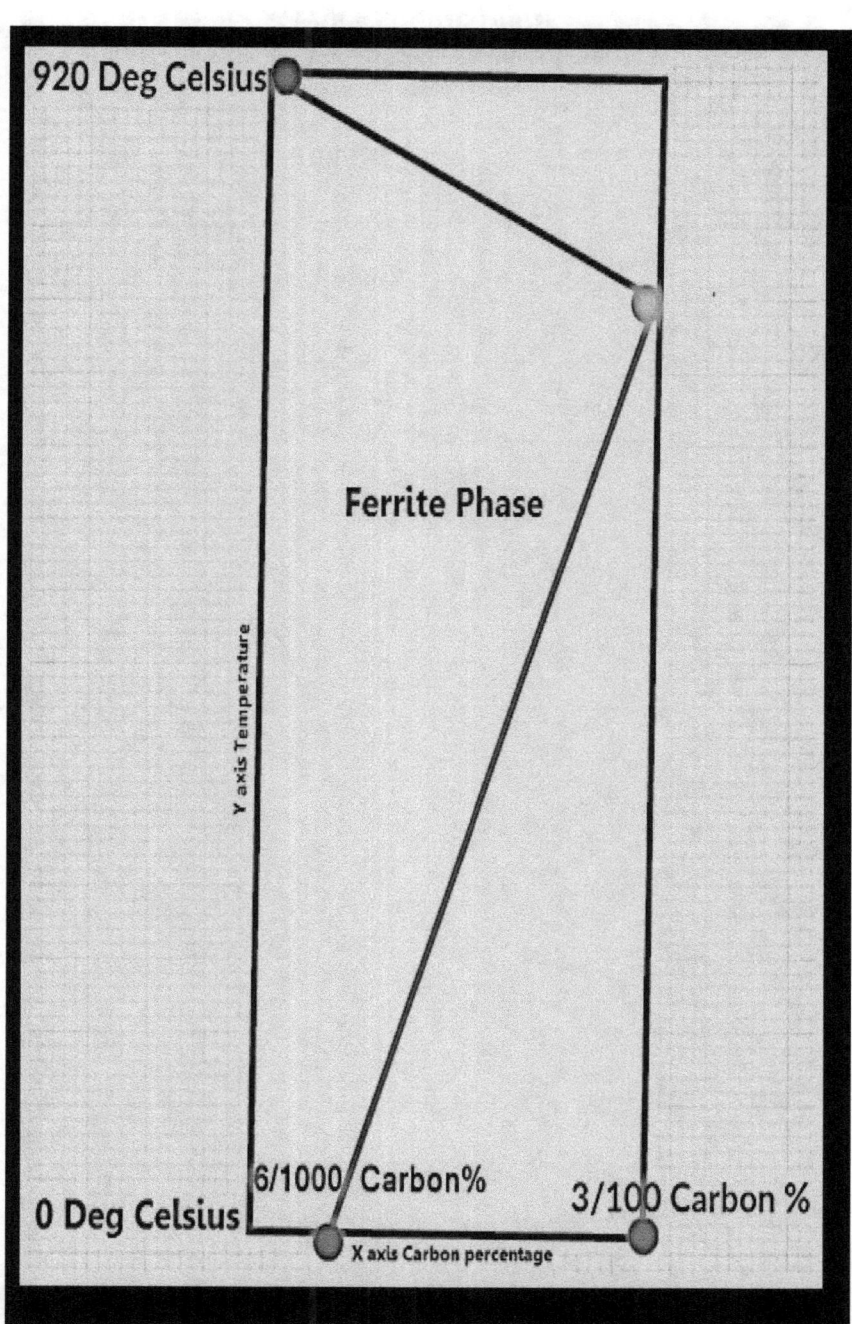

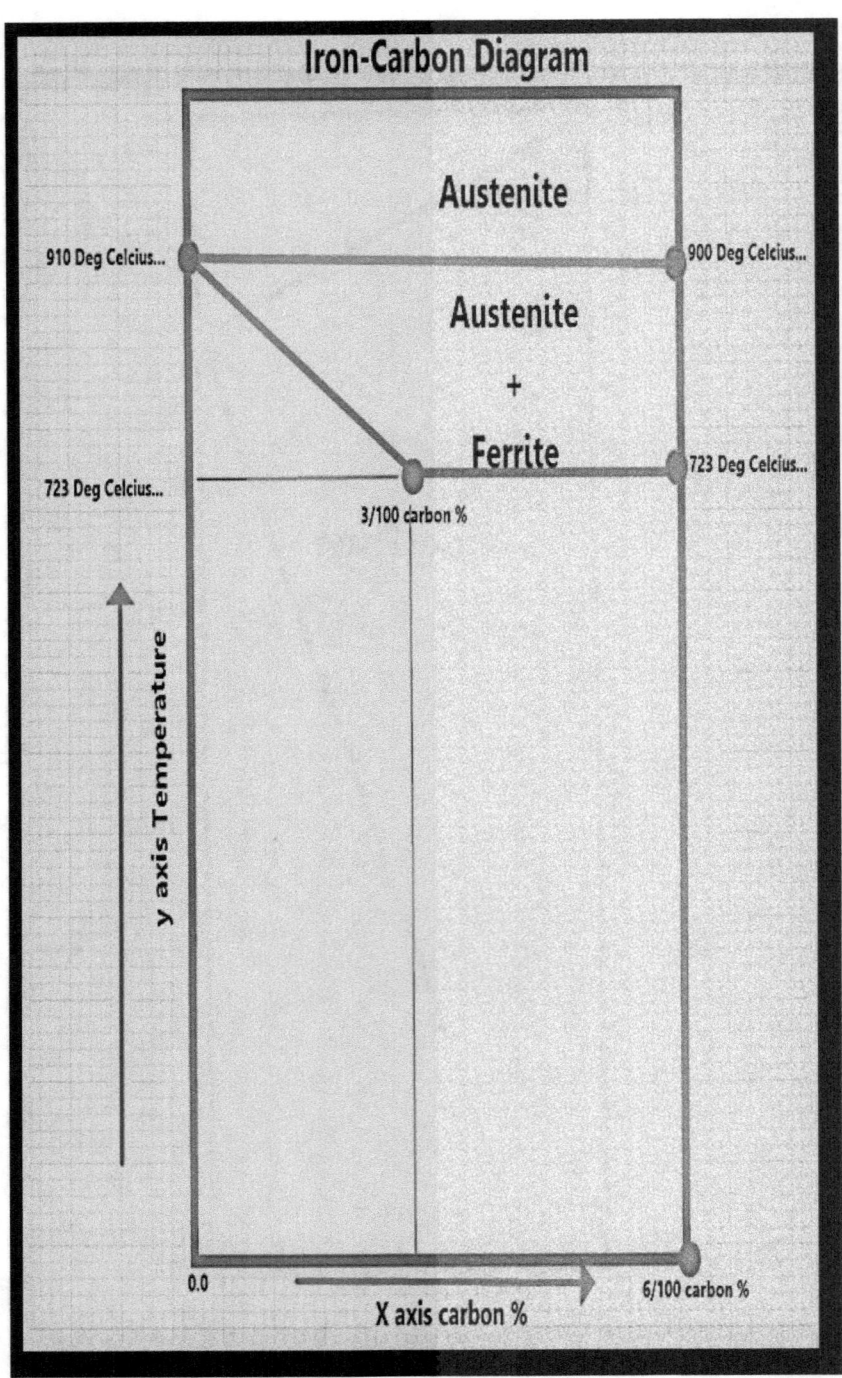

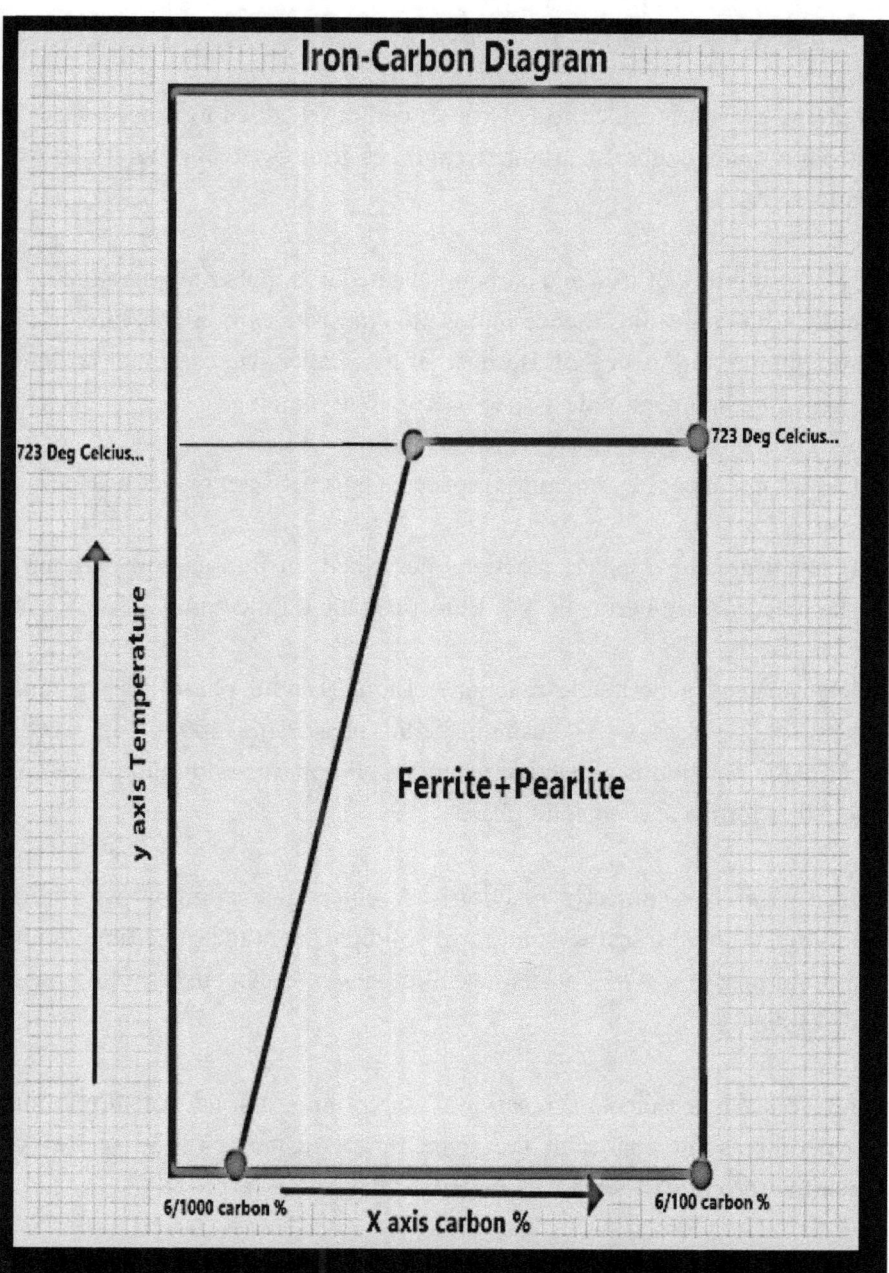

1) What will happen if Martensite is reheated?: Tempered Martensite forms upon re-heating of martensite.

2) Why carbon is added to Fe(Iron)?: carbon is added to iron to make it stronger and tougher. Carbon strengthens iron by distorting its crystal latice.

3) Why mixture of iron and carbon is heated to a higher temperature?: Fe and Carbon are solids, hence immiscible to make carbon miscible; in the iron, the mixture is heated. Because of its smaller size carbon atoms gets trapped in the interstitial3.1 space between iron atoms.

3.1) what is meant by interstitial space?: Gaps between two atoms etc.

4) What are the 3 phase reactions described, in iron-carbon diagram?: The reactions are Peritectic 4.1, Eutectic 4.2 and Eutectoid 4.3.

4.1) What is a peritectic reaction?: Liquid(Liquid phase) + gamma iron(Solid phase) ↔ austenite(Solid phase) (@1400oC) in words, peritectic reaction is a reaction where a solid phase and liquid phase will together form a second solid phase.

4.2) What is a Eutectic reaction?: A eutectic reaction in which, on cooling, a liquid transforms into two solid phases at the same time. liquid (Liquid phase) ↔ austenite(Solid phase) + cementite(Solid phase) (@1300oC).

4.3) What is a Eutectoid reaction?: On cooling, a solid transforms into two other solid phases at the same time. Austenite (Solid phase) ↔ pearlite (mixture of ferrite & cementite).

5) What is meant by metastable 12.9?: Existence of a substance as either a liquid, solid, or vapor under conditions in which it is normally unstable in that state.

6) What are the two factors mainly influences the iron-carbon reactions?: Main, factors influencing the iron-carbon reactions are carbon percentage and Temperature at which heating is carried out.

7) What are Hyper-Eutectic &Hypo-Eutectic steels?: Hypo -eutectoid steels, (ferrite + pearlite) (Carbon % between 0.022 to 0.77), Hypereutectoid steels (pearlite with network of cementite) (Carbon % > 0.77).

Chapter Six Heat treatments

1) What is heat treatment?: Heat treatment is the heating and cooling of metals to change their physical and mechanical properties. Heat treatment could be said to be a method for property alteration of metals.

2) Mention the types of Heat treatment Process?: Annealing 2.1, Normalizing 2.2, Spheroid-sing 2.3, stress relieving 2.4, Age hardening2.5, Quenching2.6 & Tempering2.7.

2.1) What is Annealing?: Annealing is a process of heating the steel slightly above the critical temperature of steel ($723^{O}C$) and allowing it to cool down very slowly. (It re-programs the material# grain growth, hence relives the stress), material becomes soft and ductile. This process requires controlled furnace cooling.

Eg : (Practical Annealing Of Steel) Full annealing2.1.A is the process of slowly raising the temperature about 50 °C above the Austenitic temperature line A3 or line ACM in the case of Hypo eutectoid steels (steels with < 0.77% Carbon) and 50 °C into the Austenite-Cementite region in the case of Hypereutectoid steels (steels with > 0.77% Carbon).

It is held at this temperature for sufficient time for all the material to transform into Austenite or Austenite-Cementite as the case may be. It is then slowly cooled at the rate of about 20 °C/hr, in a furnace to about 50 °C ,into the Ferrite-Cementite range. At this point, it can be cooled in room temperature air with natural convection.

2.1.A) Full Annealing ,2.1.B) Isothermal Annealing ,2.1.C)Process Relief Annealing 2.1.D) Spherodising Annealing,2.1. E) Isothermal Annealing.

2.2) What is Normalizing?: Heating the metal between temperatures 760°C and 990°C according to the carbon content of the material, Keeping the temperature constant according to material thickness, Slow cooling down in air.

Eg: (Practical Normalizing Of Steel) Normalizing is the process of raising the temperature to over 60 ° C (140 °F), above line A3 or line ACM fully into the Austenite range. It is held at this temperature to fully convert the structure into Austenite, and then removed form the furnace and cooled at room temperature under natural convection. This results in a grain structure of fine Pearlite with excess of Ferrite or Cementite. The resulting material is soft; the degree of softness depends on the actual ambient conditions of cooling.

2.2) What is Spheroid-sing?
Ans: Spheroid -zing is a form of heat treatment for iron-based alloys, commonly carbon steels, in order to convert them into ductile and machinable alloys. It is conducted at temperatures that are slightly below the eutectoid temperature and then it is slowly cooled.

Eg: (Practical Spheroidi- sing Of Steel) Spheroidization is an annealing process used for high carbon steels (Carbon > 0.6%) that will be machined or cold formed subsequently. This is done by one of the following ways:
1.Heat the part to a temperature just below the Ferrite-Austenite line, line A1 or below the Austenite-Cementite line, essentially below the 727 °C (1340 °F) line. Hold the temperature for a prolonged time and follow by fairly slow cooling. Or
2.Cycle multiple times between temperatures slightly above and slightly below the 727 °C (1340 °F) line, say for example between 700 and 750 °C (1292 - 1382 °F), and slow cool. Or
3.For tool and alloy steels heat to 750 to 800 °C (1382-1472 °F) and hold for several hours followed by slow cooling.

2.3) What is stress relieving heat treatment?: Stress relieving is applied to both ferrous and non-ferrous alloys and is intended to remove internal residual stresses generated by processes such as machining, cold rolling .(Treatments above 900°C are often full solution anneals) different materials have different stress reliving heat treatment temperature.

2.4) What is Aging or Precipitate hardening?: Precipitation hardening, or age hardening, is a heat treatment process that produces uniformly dispersed particles within a metal's grain structure that hinder dislocation motion, thereby strengthening the metal. The formation of these precipitates is done by using a solution treatment at high temperatures prior to a rapid cooling process.

2.5) What is Quenching?: Quenching is bringing a metal back to room temperature after heat treatment (such as annealing, normalizing or stress relieving) to prevent the cooling process from dramatically changing the metal's microstructure.(Fast cooling # Forced cooling)#Water quenching, Oil Quenching .

2.6) What is Tempering? : tempering is a process of heat treating, which is used to increase the toughness of iron-based alloys. Tempering is usually performed after hardening, to reduce some of the excess hardness, and is done by heating the metal to some temperature below the critical point for a certain period of time, then allowing it to cool in still air. Wikipedia.

Chapter Seven Misallaneous

What are the strongest chemical Bonds?: Strongest forms of chemical bond are the ionic and the covalent bonds.

Why molecules have less energy compared to individual atoms?:This difference in energy is possibly due to the fact that when atoms combine to form molecule, the attractive forces are created which result in release of energy.

What is Isothermal Heat treatment?:Austempering is an isothermal heat treatment that, when applied to ferrous materials, produces a structure that is stronger and tougher than comparable structures produced with conventional heat treatments.

Material is heated to Austenizing temperature(red hot) in a Inert atmosphere, then are quenched in a bath of molten salt at 232°C to 399°C. The quench temperature is above the Martensite starting temperature. Therefore, a different structure (not Martensite) results. In Austempered Ductile Iron and Austempered Gray Iron the structure is Ausferrite, and in steel, it is Bainite.

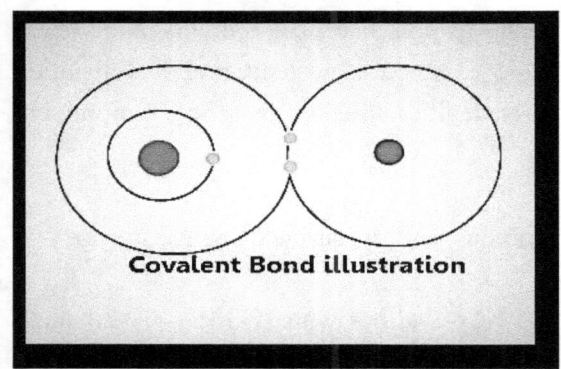
Covalent Bond illustration

1) Low-carbon steels: % wt of C < 0.3
2) Medium carbon steels: 0.3 <% wt of C < 0.6
3) High-carbon steels: % wt of C > 0.6

Low carbon steels:
Carbon present is not enough to strengthen them by heat treatment, hence are strengthened by cold work. They are easily weldable and machinable Typical applications: tin cans, automotive body components, structural shapes etc.

Medium carbon steels:
They are less ductile and stronger than low carbon steels.austenitizing, quenching and tempering done to improve properties.Hardenability is increased by adding Ni, Cr, Mo. Tempering is common heat treatment done. Used for making .gears, railway tracks, machine parts.

High carbon steels:
They are strongest and hardest of carbon steels. Tempering or Aging Done .Alloying additions – Cr, V, W, Mo .Used for making Knives, razors, hacksaw blades, high wear resistance.

Stainless steels:
They typical consists min.12% Cr . and other alloying elements, thus highly corrosion resistant owing to presence of chromium oxide.Three kinds - ferritic & hardenable Cr steels, austenitic andprecipitation hardenable (martensitic, semi-austenitic) based on presence of prominent microstructural constituent.

Stainless steels:Typical applications cutlery, surgical knives, storage tanks, domestic items
Ferritic steels are principally Fe-Cr-C alloys with 12-14%Cr. And small additions of Mo, V, Nb, Ni. Austenitic steels contain 18% Cr and 8% Ni plus minor alloying elements.

References: (References are given with due respect and credit to respective owners and authors)

1) http://education.jlab.org/qa/atom.html: Definition: Atoms.
2) http://whatis.techtarget.com/definition/proton: Definition: A proton.
3) http://whatis.techtarget.com/definition/electron: Definition: An electron.
4) http://chemed.chem.purdue.edu/genchem/topicreview/bp/ch8/ : Definition: A valence electron.
5) http://www.chemicool.com/definition/shells.html : Definition: Shells
6) https://simple.wikipedia.org/wiki/Atomic_**nucleus:** Definition: A nucleus.
7) http://www.nuclear-power.net/nuclear-power/reactor-physics/atomic-nuclear-physics/fundamental-particles/neutron/ : Definition: A Neutron.
8) http://whatis.techtarget.com/definition/molecule : A Water molecule.
9) http://education.jlab.org/qa/compound.html : Definition: A molecule.
10) http://education.jlab.org/qa/compound.html: Definition: A Compound.
11) http://www.businessdictionary.com/definition/ion.html: **Definition:** An ion.
12) https://chem.libretexts.org/Core/Organic_Chemistry/Fundamentals/Ionic_and_Covalent_Bonds : Definition: Ionic Bond.
13) https://www.britannica.com/science/covalent-bond: Definition: Covalent Bond.
14) http://study.com/academy/lesson/polar-and-nonpolar-covalent-bonds-definitions-and-examples.html : Definition: Polar bond.
15) https://www.reference.com/science/hydrogen-bonding-eee7c39b661f954d : Definition: Hydrogen Bond.
16) http://www.thebigger.com/physics/conductors-insulators-and-semi-conductors/what-is-unit-cell/ : Definition: Unit cell

17) http://www.the-warren.org/ALevelRevision/engineering/grainstructure.htm : Ans Grain in microstructure.
18) https://depts.washington.edu/matseed/mse_resources/Webpage/Metals/metals.htm : what is metal? Define.
19) https://www.thebalance.com/electrical-conductivity-in-metals-2340117 : Electrical conductivity
20) https://www.netzsch-thermal-analysis.com/en/landing-pages/definition-thermal-conductivity/ : Thermal Conductivity.
21) https://learn.sparkfun.com/tutorials/what-is-electricity/flowing-charges : Definition valence electron
22) http://whatis.techtarget.com/definition/resistance : Definition electrical resistance of a metal.
23) https://www.imetllc.com/phase-diagrams/ : Definition phases.
24) http://web2.clarkson.edu/projects/nanomat/Chapter1html/chapters/fcc.html : what is Face centered cubic?
25) http://www.sciencehq.com/chemistry/spacing-of-planes.html : Figure 7 What is face centered cubic.
26) http://www.twi-global.com/technical-knowledge/faqs/material-faqs/faq-what-are-the-microstructural-constituents-austenite-martensite-bainite-pearlite-and-ferrite/ : what is Austenite?
27) http://www.wisegeek.com/what-is-bainite.htm : What is bainite?
28) https://nayhan.wordpress.com/pearlite-martensite-austenite-dan-bainite/ what is Martensite?
29) http://www.physlink.com/education/askexperts/ae341.cfm Why carbon is added to Fe (Iron)?
30) https://www.iitk.ac.in/tkic/slides/Microstructure/L-4.pdf what are Hyper-Eutectic &Hypo-Eutectic steels?
31) https://www.decodedscience.org/chemical-bond-strongest/38154 what are the strongest chemical Bonds?
32) http://physics.stackexchange.com/questions/243582/why-has-a-molecule-less-energy-than-the-uncombined-atoms Why molecules have less energy compared to individual atoms?

33) http://www.appliedprocess.com/process : What is Isothermal Heat treatment?
34) http://www.brighthubengineering.com/manufacturing-technology/30476-what-is-heat-treatment/ : what is heat treatment?
35) http://www.brighthubengineering.com/manufacturing-technology/50523-heat-treatment-of-steels-annealing/ : What is Annealing?
36) http://collections.infocollections.org/ukedu/ru/d/Jgtz077ce/6.1.html : What is Normalizing?
37) http://www.azom.com/article.aspx?ArticleID=9708 : What is Spheroid-sing?
38) http://www.wallworkht.co.uk/content/stress_relieve_and_normalise/ : What is stress relieving heat treatment?
39) http://www.efunda.com/processes/heat_treat/softening/annealing.cfm .
40) https://en.wikipedia.org/wiki/Tempering_(metallurgy):What is tempering

CITATION
(1A)Collins English Dictionary – Complete and Unabridged, 12th Edition 2014 © HarperCollins Publishers 1991, 1994, 1998, 2000, 2003, 2006, 2007, 2009, 2011, 2014.

DISCLAIMER
Although the (I) author and publisher have made every effort to ensure that the information in this book was correct at press time, the author and publisher do not assume and hereby disclaim any liability to any party for any loss, damage, or disruption caused by errors or omissions, whether such errors or omissions result from negligence, accident, or any other cause.

www.ingramcontent.com/pod-product-compliance
Lightning Source LLC
Chambersburg PA
CBHW061235180526
45170CB00003B/1312